10|16 160 Self Sustainability
By: Ali Dawoud El

Kofi Piesie/Mossi Warrior Clan
Copyright 2020 by Kofi Piesie Research Team

Printed in the United States of America

Table of Content

Jump-off

-

By Kenneth Wayne McCrae, MSW, LCSW, CASAC, SIFI

As the Program Director of one of the largest Domestic Violence Programs in the state of New York, and I'm employed by the largest provider of domestic violence services in the Country; I witness first-hand the importance of basic needs being met.

Financial abuse is one of the most frequent reasons someone will stay in or return to an abusive relationship. My staff and I take on the challenge of empowering individuals to become self-sufficient. Too many are unable to think outside of the box and will stay or return to an abusive relationship, because their basic needs are being met.

Many of my staff who have gainful employment struggle to make ends meet, which often equates to putting food on the table, and can benefit from Ali's sharing in 10|16.

Ali Dawoud El, the author of 10|16 160 Self Sustainability has decided to address this anxiety within his own life in the event that "the paycheck was no longer there". Ali decided to face his fears and explored ways that he would be able to use what he had, to get what he need.

Ali became successful at being able to provide for his family by cultivating a small space in his backyard and dedicating it to growing and processing food.

Abraham Maslow said it best "One's only rival is one's own potentialities. One's only failure is failing to live up to one's own possibilities. In this sense, every man can be King, and must therefore be treated like a King".

Introduction

Fear is a hell of a drug. We are being systematically poisoned daily by an unhealthy diet consuming not just food but that of social media and mainstream entertainment, which contributes heavily to the dumbing down of society! Imagine for a moment the things being propped up over substance, being able to convince a significant number of our community they eat simply because they can and that the actual value of proper nourishment is inconsequential.

I've heard it said on several occasions, "it just has to make a turd" (keep in mind, "it" is referring to food). It is my position that if an adult person cannot provide **Food, Clothing, Shelter,** and the **Protection** (FCSP) of that for themselves first and then their family, they are not acting in their majority but rather in their minority. In the majority of instances, it results in having to work long hours and or having to do backbreaking work to receive, in many cases, barely making compensation; so they can then go and purchase what can be considered the resources they need to do this thing called living, all the while being exploited for their sweat equity.

Our society has devolved into the thought of if it looks good, it must be a good mentality. This thought process or mentality severely hinders a person's ability to intake the best quality nourishment to maximize an individual's potential output. The very reason we eat or need to consume nutrients is to have enough energy to efficiently complete the tasks of the day! It's a simple concept when you think about it, which may contribute to the vitalness of "it" (the importance of efficiently completing tasks) being severely downplayed. Most people I know, on any given payday, ensure to pay all their bills; before even considering paying themselves. Why invest in yourself first, you might ask? Simple, rule # 1 is safety first; if you are not safe, you are not able to assist in someone else wellbeing. Not only do they not pay themselves, but the food is usually among the last things on their mind, usually coming after the mortgage/rent, lights, water, car note, cell phone bill, credit cards/ loans, social activities, etc.

You can turn on any news station, open any social media app and see our society doing the absolute most to garner likes/attention/" nourishment" from virtual strangers. You can see citizens reacting violently towards each other in reaction to blatant unjust and abuse from members of the justice system, causing doubt to turn into mistrust and fear. You can even see the hypocrisy of those that tout piety while inciting confusion and encouraging misinformation (consciously or unconsciously). "Cancel Culture" has people glued to their devices, looking and waiting for an opportunity to participate in activities that serve them no vital benefit.

From my observations, from growing up in Brooklyn, NY, in the '80s; traveling the world on US Naval Warship in the late '90s and my time still serving my country as a civilian military member in federal service, I've been allowed to interact and exchange complex ideas and cultural knowledge with many great people. For whatever reason, I have always had a face/vibe that people felt comfortable talking to, despite how I might think about it. I accept all this, which I'm calling craziness, is due to a hormone imbalance incited by the harmful foods, images, and sounds we ingest regularly.

Look at your timeline or feed and see what is getting the most attention; better yet, log out and scroll as a guest and see what is being promoted heavily; it generally is the most debasing, impractical, foolishness you could think of. Often, I ask how do I get that time back? And the reality is that it is gone, and the content was all fear-derived at that. Fear of not having enough to provide for themselves or their families, fear of the thought of what is going to happen when you're cut off from your only means of sustenance and must fend for yourself, fear of being found unworthy by people that may or may not matter in the grand scheme of your life. All are valid, but all are also answerable!

My position is to accept what is and then create multiple action plans. There are things that I can do to improve my lifestyle and maintain the quality of life I desire to keep for myself. It is super important to think about what it means to "accept" (CONSIDERATION is vital), it is not something that should equate to being dismissible in your mind, but rather it should represent a complete comprehensive understanding of the situation; the who what when how and why's. Without ascertaining that bare minimum knowledge, you will be unprepared to apply the necessary steps for understanding to achieve a complete remedy.

Without being whole, you will continuously operate from a place of lack. You will ultimately reproduce in a state of lack and continue the cycle of just doing enough to get by—surviving as opposed to Thriving.

This brings me to why you are even reading this book or even why you should care about what it is I have to say. I'm saying support your interests, I'm saying maximize your resources; I'm saying grow your own, I'm saying learn to build your own, I'm saying learn to protect your own, I'm saying support your local farmer, encourage vocation training, form or join a team that supports your interests. Suppose any of these things are outside your skill set or ability in reality. In that case, it is your responsibility to seek out those trusted resources that'll provide it for you (just accept the dependency it creates) so you can support your interests and live.

If I were to name what I do, I'd call it Poly farming. I like the name not only because I employ a bunch of different systems and techniques (Aquaponics, In-ground growing, Container growing, raised beds, buckets, etc.), and I endeavor to produce entire meals (Chicken, Ducks, Geese, Bass, Trout, Medicinal Herbs, Spices, Vegetables, and even microgreens). I have developed an "aquaponic thought process" (wherein, I'm now thinking about things a different way: It's more of a how do the persons, places, and things I'm engaging communicate with the totality of what it is that I'm doing to accomplish my goal), and ever since building my first system, I can personally attest to just becoming a better thinker. In aquaponics, you're really trying to minimize the load on your resources, while getting optimal if not better than optimal results in yield. I have a whole thought in my head that is being realized daily, and its exciting to walk this walk. Yes, there are challenges, but the "ing" in living, implies work/action. You will be tested, disappointed, and feel like you have no talent, but if you keep the "ing" in living in mind and remember to work, it will come to fruition. All work is rewarded, sometimes negatively, but primarily positively, from what I've experienced these last 44 years. Creating full meals in relative comfort will free up mental space, allowing other aspects of your whole self to grow and be profitable.

Starting this adventure, the main things I kept in my mind for sure are all my family and friends who live in apartments or don't have the space to grow in a "traditional" sense of farming. I put quotations because I find it ironic that what is being considered traditional is an adaptation and bastardization of a tried-and-true method of self-sustainability (what I'm referring to is the commercialization or weaponization of the agriculture industry, depending on how you look at it. I've thought about all the excuses people told me to my face for having stopped them from either doing for themselves or supporting those doing what keeps their interest.

In the relatively short amount of time practicing self-sustainability, I've had the privilege to hear some of the wildest questions and ludicrous advice (from non-farmers/growers) about food cultivation; the questions themselves are not ridiculous, but the fact they have never asked these questions in all of the years they have been consciously purchasing, preparing and consuming food, is what disturbs me and lends to why I'm writing this.

My goal was to remove every excuse/reason a person has come up with over the years (not having enough lighting; not enough yard; I would, but I can't grow anything but old, etc). I've heard a lot, and now, I either want to strongly encourage a person to invest in themselves or allow for it to be easier for them to admit out loud that they didn't WANT the things they said they wanted in the first place. The reason is that once you accept what is and are willing to do what needs to be done, what will happen, you will be prepared to reap what you've sown.

1st Joint: What is 10|16 160 Self Sustainability?

"10|16: 160 Self Sustainability" is a concept that came to me in two organic ways. The latter being an NPR episode that I was listening to one day after a very long day at work. The thing about my job is for a long time it was very painful to show up every day, at one time I felt as if I was literally in bed with the enemy. Over time the negative feelings passed and was ultimately replaced with less of an emotional reaction and way more of a real reality response. My job is not my enemy; it's my plantation. I've had a million ups and downs, too many to get into here, the important take away is I was suspended without pay for two weeks, for what I would call trumped up charges. The important lesson to be learned at the time, was that it really didn't matter what was true or just (depending on the system you are under this case military/OPM); it only mattered that the powers that be made happen what they wanted to happen.

During that time, the reality of what it meant to be expendable was made quite evident. I could Not provide FCSP; I am an able-bodied man, but without a job I was unable to provide the bare necessities to live. To look in my family faces and not have a plan made me feel awful. After the two weeks of processing my emotions, and my thoughts, I remembered the NPR episode, and I remembered what my immediate responsibilities were, and I focused on creating those things in real life.

In this episode, they were discussing backyard gardening, and how a 120 sqft garden could feed a family of four for an entire year. Hearing that got me to thinking about what was being asked (10 ft x 12 ft) and if I had it, turns out I had that with some to spare. It's crazy because at that same time I was heavy into my ancient ancestral studies, I was a Kmt and West African freak. I loved everything African; to me African was synonymous with being black at the time.

I had been travelling the world from 17 years old, and I was fortunate to have port visits in many historical sites and countries. During one of my studies, I was shown an image from Kmt that depicted what looked like a lake filled with fish, surrounded by vegetation. At this exact time, to me it looked like a brand-new system (to me anyway) I had just heard about, "Aquaponics". Immediately because it was African, I was all in, I knew right then and there the first step towards self-sustainability was for me to build an aquaponic system. I completely forgot about NPR and went with the African thought, it made more sense, I could kill two birds with one stone so to speak.

The first organic way in which 10|16: 160 Self-Sustainability has been foundational in my life is my upbringing. Growing up during just one of another devastating acts of mental genocide in this country, I'm referring to the crack epidemic in the in the heart of Brooklyn, mid 80's – mid 90's. I learned early to hold on to what you got, have everything in its most functional space by keeping it close and utilizing all your space. If you've ever had to hide money in different parts of your outfit in order to only lose the least of your resources vice the totality, you know what I'm saying. You know what it means to survive in a literal sense.

At home I learned the value of not limiting my options and to utilize my corners and vertical space from my old Earth (my mother). By taking our tiny Brooklyn apartment and making it everything we needed. I cannot think of a single instance when I thought we were poor growing up, we moved furniture so often, painted walls, we changed our perspective it seems like every six months. I probably could've gone into interior decorating had I been paying attention. I could've been on HGTV or something with one of those shows everybody watch and attempt to imitate

Having learned the harsh lessons of growing up surrounded by lack raised in the crack era, crazy violence, and hookers galore, I took what they said on NPR and combined it with what I was doing; and expanded on my aquaponic system and came up with a scalable system of sustainability that I hope you put to the test and see if I'm right. By no stretch of the imagination is *this* simply a theory but an absolute fact

If you endeavor to go on this journey, this guide will serve you well in your efforts. Anybody that can attest to knowing who I am would tell you that if I'm interested in something and you ask me a question, you might want to be sure you have the time and desire to learn. I tend to run my mouth and tell everything, most of the time, too much if I'm being honest. One evening while building with the gods (my brothers), I was asked how to keep enough chickens to provide for a specific size household, it being essential to be: low maintenance in terms of general care and upkeep, low on the noisiness, and doesn't stop you from going on vacations.

So, I rattled off a solid solution, and what do you guess happened next? Like it happens too often, the answer was met with illogical criticism and resistance. (I'm the type of person to consider these types of thought processes as dangerous and would cause a need to avoid these individuals at all costs; BUT when exceptions are made, you must accept and adjust). Luckily it was a panel discussion because one of the other gods blurted out, "Yo! you just dropped a million dollars' worth of game, god", the god told our brother. "He just answered all your questions and concerns, and you're still looking for ways not to do it."

My man Pearl then added on and asked about his situation and setup and what he could do. I considered what he asked, and I looked out of my window, and I said you know, with what you have, you can have six chickens with his current set up, but if he considered enlarging his enclosure to a 10x10, he would be able to keep 50 chickens. Before I could finish my thought, he exploded in excitement, saying, "that's your book, god!" and here we are.

To finish my previous thought in totality, my brother Tahir was told he could get an X-large dog cage and keep three hens comfortably in his living room. To me, this could be like owning a parrot or any other exotic type of bird. The thing about this exotic (not too many people keep chickens) bird is that it is beautiful and provides nourishment that contributes to your ability to complete tasks. I told him that if you were to rap the cage from the outside (securing all entry and exits) with chicken wire and whatnot, he would be able to contain the bedding and feed from when the hens scratch; he would also ensure no escapees, resulting in chicken poop everywhere.

- This setup might seem farfetched, but it consists of:

- a dog cage (about $200 online, or repurpose an old one),

- chicken wire (about $45 for 25ft) Hardwire cloth (metal chicken wire)

- wood shaving (about $-10, depending on where you purchase, feed stores are cheaper than pet shops),

- a nesting box, or a bucket (don't overthink it)

- feeder (bowl $5 or automatic variable in cost but can be built for about $15),

- waterer (bowl $5 or automatic variable in price but can be built for about $10),

- A perch of some kind (1" PVC pipe is good and stiff enough to support the weight of the birds, for example)

- A heat light for the early chick-rearing, better known as "brooding" days, about two weeks

- Chicken feed price is variable, but a 50lb bag of organic feed usually costs about $35, but leftovers from what you ate for dinner also counts as chicken feed

- Finally, you need chickens.

With the above setup, you can easily keep 3 or 4 birds, depending on the breed, for a solid year and a half of consistent laying time. Because this is an indoor setup, the elements in nature are non-consequential for the most part. If you're wondering about the smell, because I know chickens smell right, with enough wood shavings as bedding, a process called "Deep litter" reduces if not altogether eliminates the odor, coupled with Diatomaceous Earth (DE) and Lime you'll keep the smell and pests down severely. Keep the litter dry, and most of your problems are reduced drastically. The thing about deep litter is you add bedding instead of scooping it out weekly or daily, depending on how messy your girls are. Notice I said girls, roosters Crow whenever they want, and they are not an in-house animal.

Pearl had a large kennel already, which is why he was so excited. His enclosure measured 5ft x 8ft. With the average poultry farmer utilizing a 2-3 sq ft per bird ratio, his setup could easily hold 6 – 8 birds (for egg laying) after considering the space the amenities take up. When I was telling Pearl about his potential, I happened to glance to the other side of my yard, looked at one of my setups, and suggested that if he were willing to expand just a bit, he could get a 10 x 10 rap it. He could easily hold up to 50 meat birds (birds that grow to harvest weight in 6-8 weeks) in that area. And if he wanted to keep going, with about another 6ft adjacent to it, he could build an aquaponic system, where he could cultivate his fish and his veggies simultaneously. I said with a 160 sq ft space (1016: 10 sq ft x 16 sq ft), you could grow all your chicken, fish, and veggies for yourself and your family in a reasonable amount of time.

And that's the blueprint. With a 10 x 16 space, you will be able to:

- Raise enough chicken to have chicken every week in any given year in about two months increments

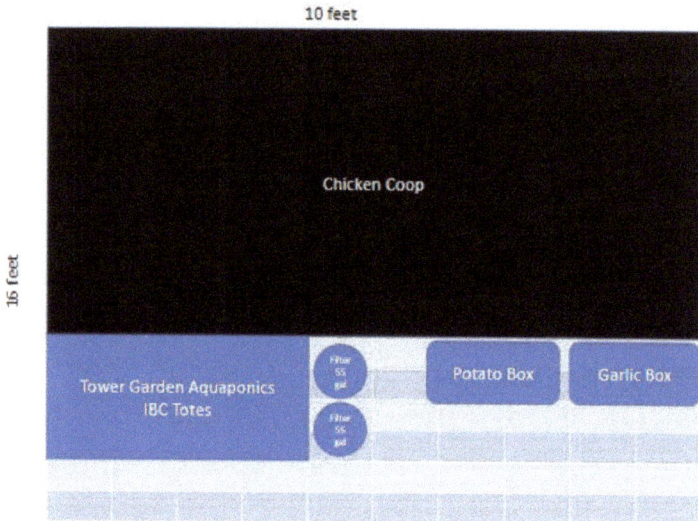

- Raise enough fish to reduce your seafood costs significantly in any given year, from about nine months to a year and a half, depending on the fish

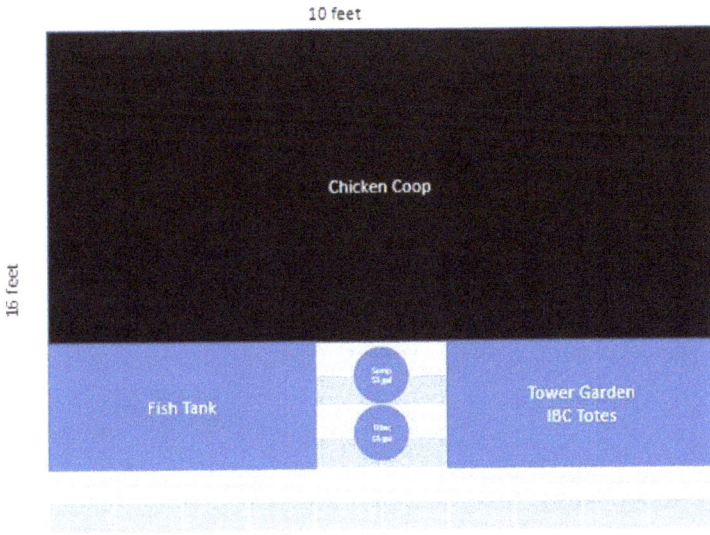

- Grow perennials and standard herbs and spices you use daily

- Grow seasonal vegetables

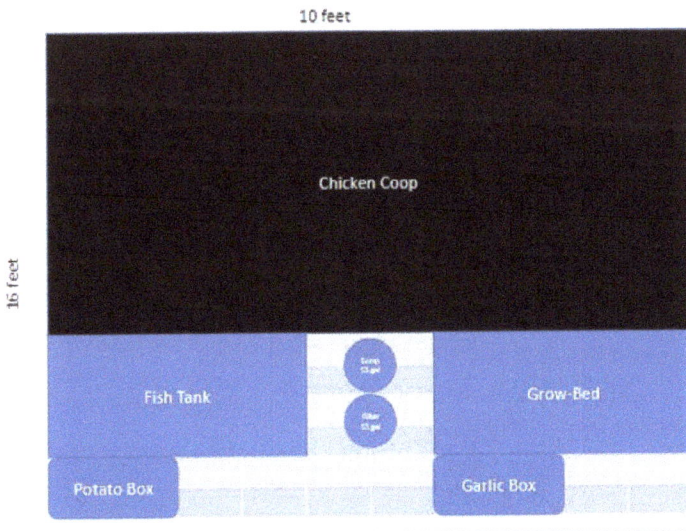

- After the first two months and you harvest your flock, you can raise another 50 birds, and you'll have extras for when you want a bag of wings or leg quarters or whatever (you can do this about three good times a year, allowing some time for yourself to reset)

- designate an area for composting: when you turn or discard the old bedding, you can use it to raise black soldier fly larvae or any larvae of your choice; this will significantly reduce your chicken feed cost, as they love grubs and will produce excellent fertilizer that can be sold or put in a potted or raised bed garden

I can guess what you might think; this cannot be real, right? There must be a catch; well, you'd be correct. The catch is, do you really want the things you say you do, or do you just want to chill? That is the question you must answer truthfully for yourself. With this book, it's laid out in front of your face, and you must be the one to decide if the stomachache is worth the bite.

I'm a problem solver. I think it is dangerous for anyone to action themselves from a place of emotion; thus far, my conclusion is that critical thinking (the understanding of the mind) is essential to solving problems efficiently and effectively.

It consists of:

- 10 x 10 Kennel or canopy

- Hard Wire cloth

- Chicken wire

- wood shaving (about $-10, depending on where you purchase, feed stores are cheaper than pet shops),

- feeder (Large bowl/flat long container (like a roof gutter) or automatic variable in cost but can be built for about $15+),

- waterer (55-gallon barrel $45, or commercial automatic $$$$)

- heat lights for the early chick-rearing, better known as "brooding" days, about two weeks

- Chicken feed price is variable, but a 50lb bag of organic feed usually costs about $35, but leftovers from what you ate for dinner also counts as chicken feed

- Finally, you need chickens.

- Fish Enclosure (IBC totes $99, Rubbermaid stocking ponds $125+ depending on size, build concrete pond Costs depend on design, PVC pond pool liner costs depend on the method, but PVC is like $15 for 10ft, etc.)

- Aquaponic grow beds (IBC totes, Juice Bottles, etc.)

There you have it, the complete what you'll need to establish a sustainable lifestyle.

2nd Joint:10|16 160 Self Sustainability: Aquaponics

What is aquaponics?

Using a simple google search, one typical may find an answer to what aquaponics would be; integration of a hydroponic plant-producing system coupled with a recirculating aquaculture system (RAS). I know that didn't answer the question, right? So, the first step is to accept that aquaponics is a hybrid system, a combination of two separate forms of cultivation.

One is hydroponics, wherein the idea is to grow plants without the use of soil. Any artificial media will do once you get into the mode of experimentation. I've had successful **growth** using clay balls, lava rocks, cotton balls, and a few other things as some examples.

This method is suitable, but you must know that you'll always need nutrients from somewhere. This can be done quickly by purchasing the most common nutrient, fish waste. Or my preferred method, you can take it a step further and set up a small fish tank and use the water from when you do your aquarium maintenance. You could use your dirty fish tank water as your nutrients, and it would cost almost nothing to feed a bunch of gold feeder fish. You could also set up a separate solid waste container designed to aerate and agitate waste over time, making it the perfect plant fertilizer. You would have to ensure you have enough bacteria growing; otherwise, it will take forever to be helpful.

The companion system that makes up aquaponics is called "aquaculture," or RAS, wherein instead of plant cultivation, you'll be raising fish for consumption, pet shop, or commercial use. RAS are normally closed systems where the water is maintained using a set of filters (i.e., biofilter, solid waste, sand, charcoal, rocks, etc.). I like this idea because just about everybody I know has either had a fish tank in the past or has one currently. To me, this means instead of those pretty, expensive, show-off fish, you could substitute them for food fish. The one negative about this method of cultivating is the wastewater. While it's minimum due to the use of the filter setups, there does come the point where you would have to change the water. It might seem like a simple problem, but simply dumping this concentrated amount of waste in the ground or local waterway may negatively affect the environment, depending on how heavy the solids are.

When you Combine these two independent thought processes, you create an entirely different thing, called aquaponics. Solid fish waste is removed, and the ammonia in the water is converted into nitrites. Microbe (interestingly enough, you need bacteria, you want it, you'll do the most to promote the growth of bacteria, I promise you) cultivated in your grow media and filter media will turn into nitrates. Plants naturally remove nitrogen from the water; this creates a symbiotic relationship (one hand washes the other) with the fish, so when the water returns to the fish tank, it is now pure and high in oxygen.

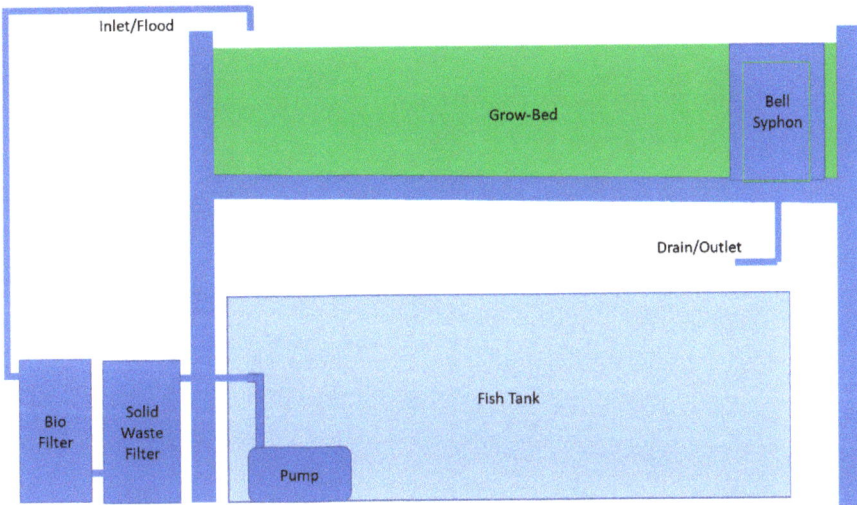

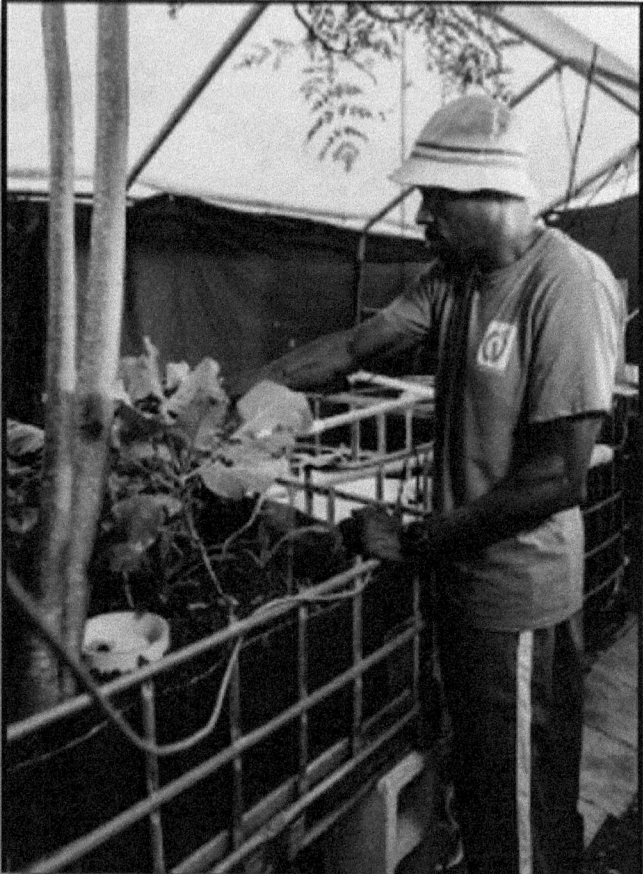

What is the benefit of aquaponics?

The most significant benefit I've seen of aquaponics is how it forces you to think about your situation in a methodological way. If there is a functionality issue, you break down the components in your head in terms of what they should do. Once you decide to act, ensure that when you clear the cause after troubleshooting, you're now sure that each component is operating as designed, even if only in your head.

Another massive benefit to this growing method is the water you save. Once you ensure you have no leaks in your piping, the only obstacle left is something you cannot do anything about, the sun. Evaporation is natural, but luckily this will only cause you to top off your water about once a month.

Another benefit of growing aquaponically is the lack of weeds present in your system. Like most raised garden beds, it's rare to cultivate weeds. From my experience in my in-ground beds, weeds spread from general lawn care, and it's tough for me to cut my grass and not disperse spores everywhere to include in the garden. It doesn't help that I hate mowing grass, so maybe I'm just doing it wrong, but whatever, this is a limiter, if not an all-out elimination of weeds due to the system's height off the ground. Most IBC totes are about 4 ft high, having 4 ft being about your base height of the overall design.

I might have forgotten that this system is primarily gravity fed. If you could figure a way not to need a pump, then you would be entirely off the grid and significantly reduce the need for electricity for a pump. I'm not that smart, so I rely on the handy dandy pump and let gravity do the rest.

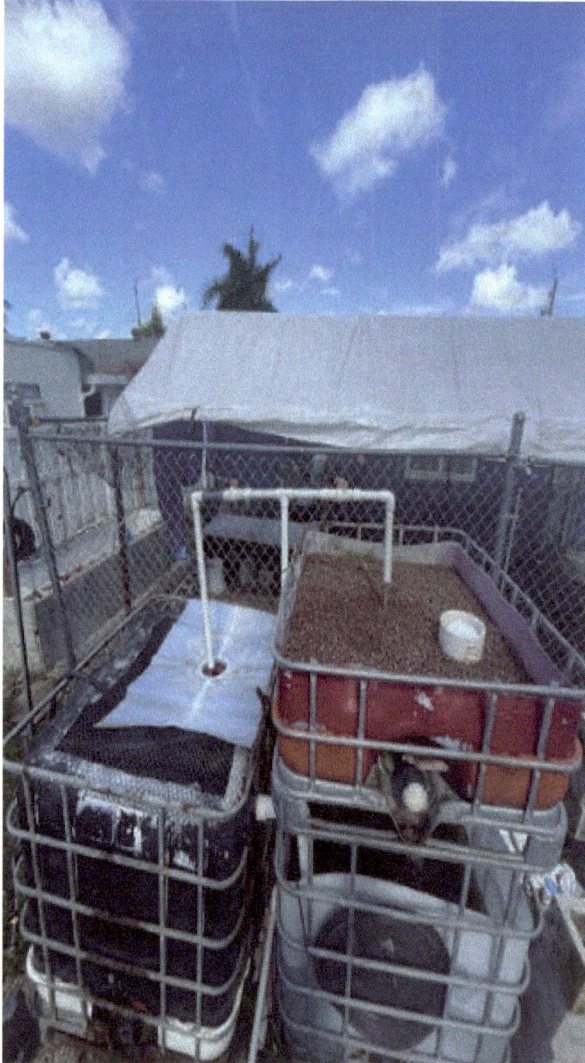

The Tower System

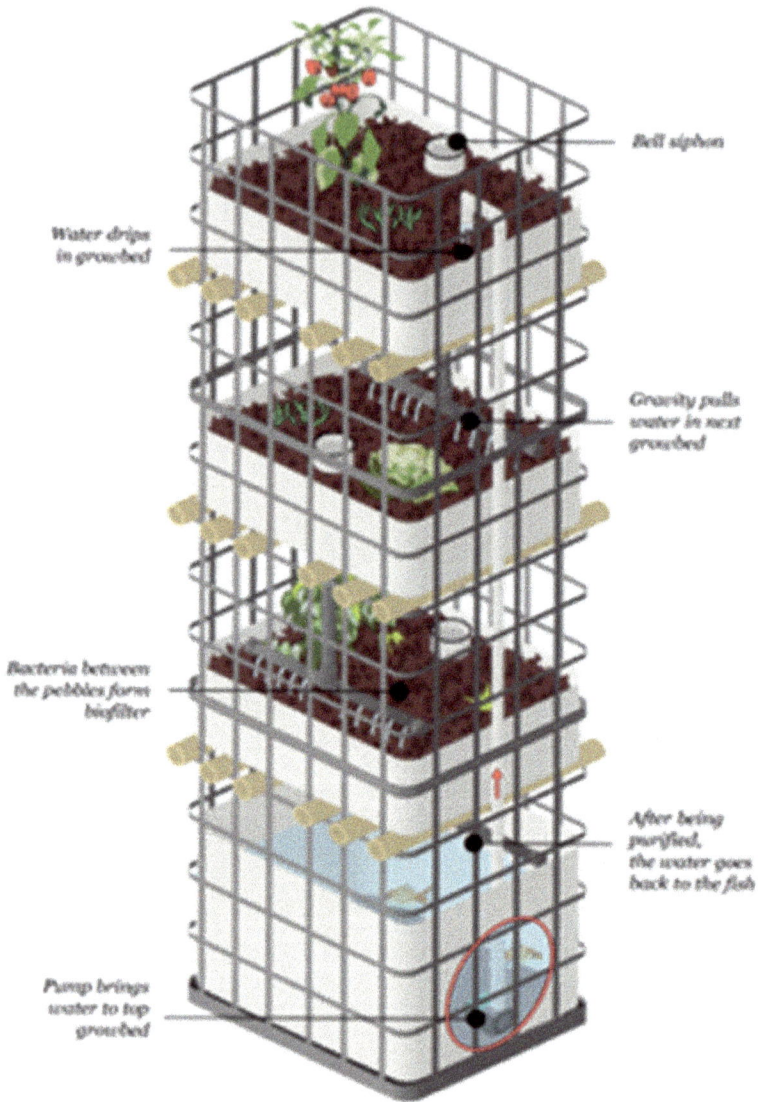

Bell siphon

Water drips
in growbed

Gravity pulls
water in next
growbed

Bacteria between
the pebbles form
biofilter

After being
purified,
the water goes
back to the fish

Pump brings
water to top
growbed

Els Engel

What are the drawbacks of aquaponics?

The number one drawback in growing aquaponically is your nutrient balance, so the truth is fish and plants desire and require different levels. Depending on what plant and animal species you choose to cultivate, something as basic as PH will determine what levels you need to maintain.

This is why I look for local fish; whatever is growing around or near me is what I want to grow mainly. They're pretty tough and durable from what I can see (swimming in the canal out back, and I can see how filthy that water is). So, I like catfish, Tilapia, largemouth bass, and any kind of sunfish. These fish are sturdy and can withstand the clumsiness of my farming style. They tolerate almost any situation, so pairing them with plants is easy. Now, if you want something super exotic, I'm not saying don't do it. What I am saying is to do your research first. See what PH level that species thrive in and find the plants in your zone that match that fish. It could be a natural or straightforward concept; it's up to you.

Adding nutrients can be tricky; my experience suggests that you should never add anything directly to your fish tank; either add it to your sump tank or one of your filters. I've killed more fish and plants than a little bit because I put too much (not following recommended dosing) or directly into my fish tank.

The key to this methodology, or do anything really, is to take your time, be patient and let the process work. Once you've seen the entire process, then and only then can you make any accurate adjustments. You cannot fear a part of the process not working. It must be OK to fail, and it must be OK to start over. Again, it is my experience that failing and not giving up makes you a more intelligent person and more reliable.

Components of an Aquaponic System:

- Pump (1 typically needs to be powerful enough to reach your highest point in your system and circulate the amount of water collected in your sump)

- A sump tank (which needs to be able to hold enough water to maintain the water level in your fish tank and fill your grow beds) is used as the recycled water source for your plants and animals. Without it, the levels in your fish tank would rise and fall every time they grow beds filled and drained. This will cause a lot of stress on your fish if it's drastic.

- Grow bed (depends on how many you want

- Fish tank (just needs to be able to hold water without losing its form and not corrode)

- Filter container (just need to be able to hold water without losing its form and not rust)

- Piping (PVC or tubing, depending on your preference)

- Bulkhead fittings (depends on the size pipe or tubing you choose)

- Bell Siphon (you can make one or buy a kit. I prefer this as it is simple to understand and troubleshoot)

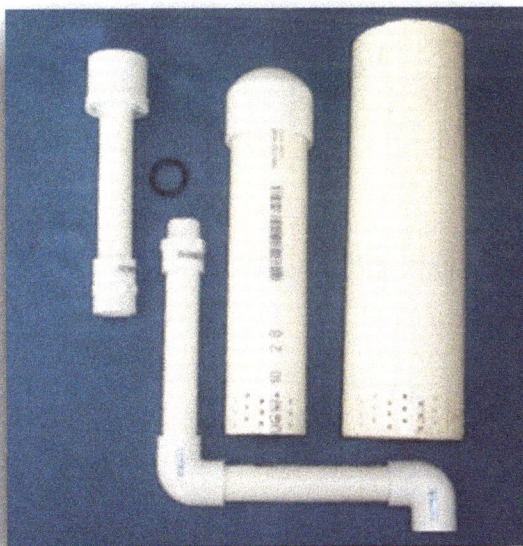

- Grow media (clay balls, cotton balls, rocks, sand, etc.)

- Filter media (lava rocks, charcoal, cloth, sand, etc.) almost anything can be used to catch debris

- Aerator (to add oxygen to your system. Your fish need distilled oxygen to survive, without it, the water will be too toxic)

- Nutrients

You have the list of components; this is where it gets interesting. How you envision your system design dictates how much piping you'll need. I encourage you to draw out your design ideas on paper, look at your space and trace it in the air with your finger where you anticipate it going to see if it still fits. If you're happy with your drawing and have done your measurements, get to the building!

How has thinking aquaponically affected my lifestyle?

The pandemic is an obvious example of how this process has shown up in my reality. I was reassured in knowing that my family and I were super comfortable. We've built or installed just about everything we felt we needed to sustain ourselves a while back. My son and I made a gym to include a pullup bar and dip bar; I have food growing around my house. I had enough things in my immediate vicinity to keep our minds occupied during the lockdown. Unlike many other people, I had no natural urge to have to leave my house to be stimulated.

Types of Aquaponic Systems:

There are generally three types of systems you can employ. There is no right or wrong answer to solve your specific problems. I will discuss the first system called Nutrient Film Technique (NFT). This system is a "Hydroponic" grow method, but because we focus on incorporating fish into the mix, it becomes aquaponics. This system maintains a shallow water supply level to keep the plants' bare roots constantly semi-submerged in nutrient-rich water. Imagine the process of germinating fruit seeds using a jar of water and toothpicks. The idea is not to submerge the basis in water but to make the water close enough for the plant roots to reach and grow towards the nutrient source.

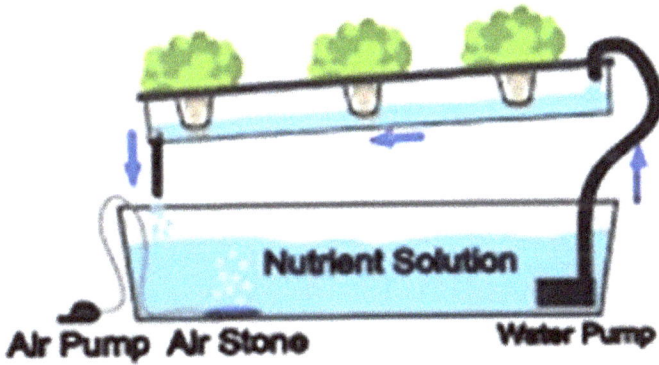

Nutrient Film Technique

The second type of aquaponic system is called the Media Bed or Flood and Drain system. This is my preferred method as it is the most hands-on. I know "the most hands-on" doesn't sound optimum, but I promise this process will make you a better thinker overall. With this system, you'll employ either clay balls or some type of rock as your soil replacement. The roots of the plants need to be able to penetrate and move freely in the non-soil.

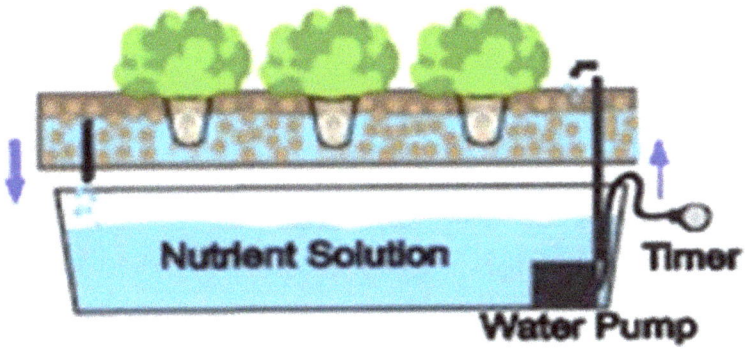

Ebb And Flow

The third type of system is called Deep Water Culture (DWC) or Raft System. This system completely submerges the plant's roots in nutrient-rich water. This system is also commonly used in hydroponics, but again, because we are multi-thinkers, we'll add fish and make it way better. The easiest way to think about this setup is by imagining a fish tank full of fish, with a piece of Styrofoam sitting on top of the tank like a lid, with whole cutouts for your planters. Again, with this, you can have any size fish tank, go to your local craft store, and get a sheet of Styrofoam, which should cost about $5. For any of these grow methods, something as simple as disposable plastic cups with holes can be used as a planter.

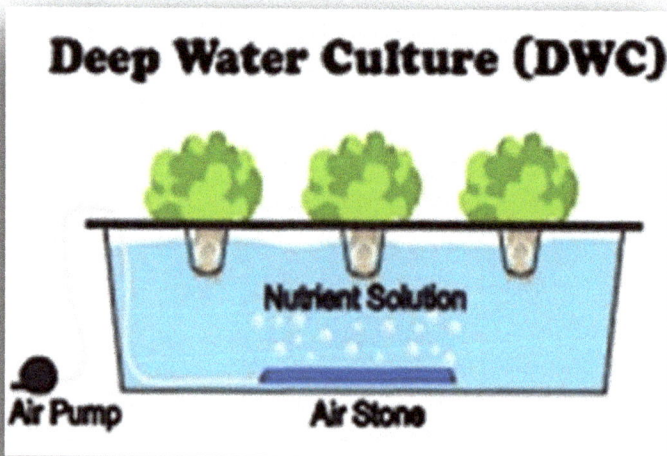

Deep Water Culture (DWC)

Each one of these methods is designed to maximize output while limiting the number of resources spent producing. By resources, I'm talking about money, time, and energy. All of these are vital; without them, life would be almost impossible.

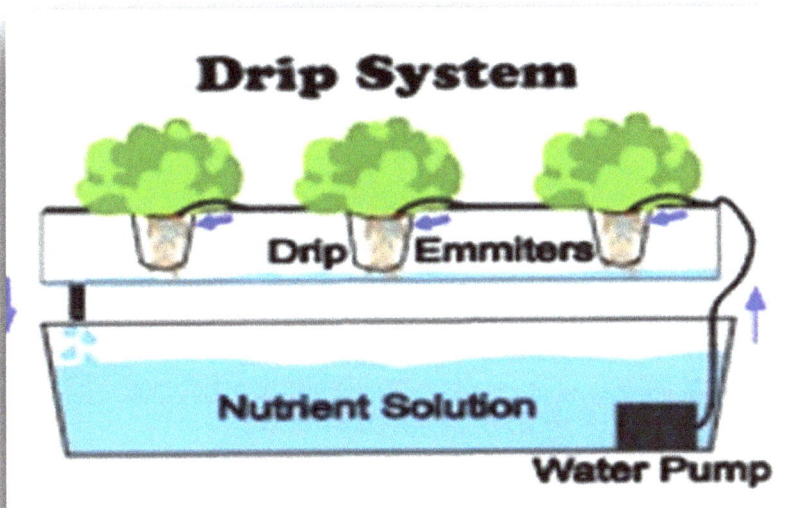

Drip System

Drip Emmiters

Nutrient Solution

Water Pump

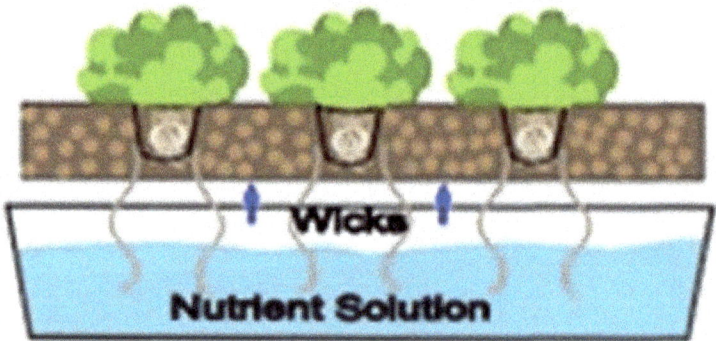

Wick System

Wicks

Nutrient Solution

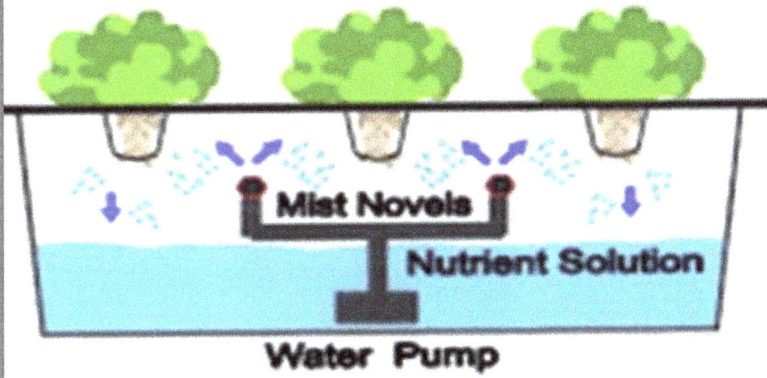

Aeroponics

Mist Novels

Nutrient Solution

Water Pump

3rd Joint: 10|16 160 Self Sustainability: Chicken Husbandry

What is Husbandry:

Poultry Husbandry simply provides an environment that supports a long and happy life for your flock. It's thinking about the basic requirements for life (food, clothing, and shelter), putting yourself in the animal's position, and asking what could make this life better. If 1000 thread count sheets are what makes you sleep the easiest, I'd ask you to consider that same level of comfort for your animals. Maybe not 1000-count sheets, but surely the bedding you choose will remain dry, plentiful, and pest-free.

Type of poultry farming:

Layers:

Laying hens require a 2–3-week brood time, isolated from danger and outside temperature fluctuations. Preferably housed inside until their feathers grow in, making them ready for the world. Interesting how something as simple as going from a fuzzy wuzzy to a winged creature drastically improves the survivability of the species. In my mind, it's akin to humans going through puberty; once they start to grow hair, it's a signal to the universe that this being is being prepared to further the population and will ultimately have to adapt to the environmental changes and challenges to continue life as we understand it to be.

After the 3rd week of brooding, you can place
these pullets (young hens under a year of age)
keep them inside the coop, where they can live
the rest of their lives. It takes about a good six
months for any bird to begin laying. The
average time hens start laying eggs is about
21-23 weeks. The first couple of eggs to be
dropped will be what I call starter eggs. These
eggs are usually tiny and malformed, and they
are excellent to consume; however, to get the
eggs you are accustomed to purchasing, I've

found that the 7th month is what I call the "golden age." At this time, your pullet has gone through the process of internal egg production, nesting, and all the social connotations that come with that. Did you know hens will share a nesting box? It's like a survival mechanism, a safety in-numbers type situation. Sharing nesting boxes can create a bit of drama in your coop; you will probably hear a bunch of cackling (two or more hens arguing) sounds that can get annoying but luckily, it doesn't persist long. Once the birds establish a pecking order (an actual order determined by height and weight), you'll discover that the noise will be significantly reduced.

Layers I recommend and why:

Black Australorp: Heat and Cold Tolerant, L/XL Brown Eggs, Laying/processing Maturity 15- 20 weeks.

Rhode Island Red: Heat and Cold Tolerant, L/XL Brown eggs, Laying/processing Maturity 16 – 20 week.
Golden Buff: Heat and Cold Tolerant, L/XL Brown eggs, Laying Maturity 17 – 20 weeks

Green Queen: M/L Green eggs, Laying/
Processing Maturity 16 – 20 weeks

Easter Egger: Heat and Cold tolerant, L/XL
Green/Blue eggs, Laying/processing Maturity
16 – 20 weeks

Meat Birds:

Meat birds are defined as birds that grow to harvest weight in about 6-8 weeks. These birds were bred to consume large amounts of feed that are then converted into fat or body mass for us to enjoy with our families and friends. Typical commercial Meat bird farmers keep their birds confined in cages that provide the bird with about 2-3 feet of space for its entire life. This containment serves to ensure the fattiest meat able to be produced. So, what I've noticed in my yard, my meat birds, even though they get processed (harvested) usually at about ten weeks, have minimal fat deposits, the birds are hefty, but because of the space (10 x 10) they have the freedom to run, jump and flap their wings. My birds are athletes. It does make the meat a bit tougher, having less fat and more muscle, but if you boil the chicken before preparing it, you will experience one of the best chickens you've ever tasted.

Meat Birds I recommend and why:

- White Cornish Cross Broilers: Processing Age 6 – 9 weeks, 5 – 8 lbs. of meat

- Broad Breasted Turkeys: Processing Age 16 – 24 weeks, 25 – 45 lbs. of meat

- Rainbow Ranger Broiler: Processing Age 9 – 11 weeks, 3 -4 lbs. of meat

Dual Purpose:

Dual-purpose birds are your everyday regular chicken, and the breed doesn't matter. I say this because these birds if spared from harvest and allowed to lay eggs are prolific egg layers and will be very profitable. But if you so choose to harvest these birds, the right time would be at the 20-week mark before they drop their first egg. If I haven't said it yet, there is no right or wrong choice; it's about what you need specifically at the time.

Brooding:

Brooding speaks to a specific period in a bird's life, either egg hatching or preparing to hatch eggs. The early stages of a chick's life can be tricky, everything is dangerous, and they are naturally curious and explorative. It's like caring for a baby that just started to crawl; everything they can see is a goal they can reach; it could be nerve-racking as a parent. It takes about 3 – 4 weeks to brood a chick completely; during this time, they are fuzzy, soft-boned, not temperature regulated, and vulnerable to everything.

For a hen to go, "broody" means she is ready to be a mom. This is a good thing and a bad thing depending on your setup. Her natural desire to reproduce will occur if a rooster is present or not, hens naturally lay eggs even if not fertilized. A hen that goes broody will severely limit their eating and drinking; they are compelled to "nest" (incubate its egg) until the egg hatches. If you are set up with a rooster, there is a good chance between 21-28 days, and your hen will be a proud new mom. She will shelter her young under her wings, ensuring they stay out of danger and don't freeze. I've watched my Ayam Cemani (All Black, skin, meat, and bones) hen hide her chicks, watched her coach them to the roosts, I've seen her carry a chick on her back to the top of the coop as if she were giving step by step instructions on the way.

While it may take up to a month for an egg to complete its incubation and hatch, a broody hen without the presence of a rooster will sit on her or any egg indefinitely. This is where it becomes tricky because during this time, not only is her nutrition intake limited, which could cause her death, but more importantly, you are now without your nutrition/profit from her daily egg delivery.

Natural brooding:

Simply put, it's a hen hatching hers and others' eggs and rearing them for weeks until they are big enough to handle their own business. The motherly instinct is strong and well expressed when you sit and watch these animals. A mother hen will spread her wings and literally wing tip to wing tip to ensure all her chicks are within her grasps, keeping them warm at night and safe from prowling predators. She will even squat for long periods not to squash the chicks taking shelter beneath her.

When a hen is defending herself and her chicks from predators, their feathers stand up big as if a blow dryer was in the background, making them look like little tanks. They charged forward and began their onslaught, usually with a peck followed by a swift talon strike. It could be exciting to watch if you're interested in how natural nature expresses itself in real time.

Artificial Brooding:

This process is more hands-on and scientific. Artificial brooding is when you take an egg from day 1 to day 28 of its life using an incubator. After this, you'll need to consider a heat source for both day and night. You'll need an enclosure to house your chicks. I prefer stock tanks; they are deep enough, so the chicks cannot fly out, nor can predators be able to walk by and see them. Truthfully you can use whatever you want; all you must do is remember that whatever you choose needs to be able to house these birds comfortably and securely. I've seen some farmers use cardboard and build an old-school fort as their brooder.

Yes, the house for your chicks is called a brooder; what you are doing is pretending to be that mama bird that spreads her wings far and wide, squatting for hours at a time, willing to puff up as big as possible and defend them from the worst of the worst. Artificial brooding means just that, you are creating a brooding situation that is naturally occurring in nature, but you're doing it "manmade style." You'll need bedding, good ventilation, and the proper lighting.

Litter Management:

Do you know how the old folks used to say cleanliness is next to godliness? I could remember saying 'yeah, right" in my head when

I was very young. But what I find is that it's kind of the absolute truth. There are so many diseases that will literally pop up out of the ether where filth is present.

Rule number one in poultry husbandry is "keep the litter dry." Just do it; it will make your life so much easier. In terms of brooding chicks, I prefer hay over large pine shavings. If you cannot find hay, don't get the fine shavings as they resemble chicken feed.

It is always best to use fresh bedding when starting chicks in their new temporary home, but if this is a cyclical thing for you, feel assured that you can reuse old bedding for new chicks. If you decide to recycle bedding, you must go through some steps to allow for the ammonia to be released. If the new chicks inhale the old chick's waste, it could devastate your flock. Literally a deathbed. Nothing worse than seeing a bunch of dead baby birds and doing your investigation and being your culprit.

The overall point of litter management is to provide comfort for your chicks (Newly hatched chicks require at least four inches of bedding to stay off the cold coming from the floor or the ground.) and to reduce the amount of ammonia in the enclosure from the bird waste.

To ensure your chicks are growing at safe rates, you'll need to have a thermometer and ensure you are getting 90 degrees in the day and 80 at night. Remember that you don't want to cook your birds alive, so be sure to have shade areas or a side where they can escape the heat if needed.

Air Quality:

High ammonia (NH3) or carbon dioxide (CO2) levels will impact your bird's health and overall growth. The good rule of thumb I've gone by is that if you can smell it, it's time to fix it. Humidity is tricky and something most people wouldn't think about with food production. Just think about a hot, humid day and how it makes you feel. Think about all the steps or measures you could take to relieve yourself of this discomfort. Now imagine that bird in those same conditions, without any way to remedy the situation on its own. It's relying on you to ensure that it has breathable air.

Sounds dramatic, right, but the truth is, I've killed hundreds of birds, not knowing the silent dangers, the things that cannot be seen or smelled that have devastated my flock in the past and possibly yours in the future. Think about this; Carbon dioxide is odorless and colorless; it takes a very long time for humans to show symptoms after exposure, let alone notice anything funny. Do you know how you can taste bad smells sometimes? Is it just me? But for chicks, when it comes to air quality, if you could taste it, the bird is probably blind and ready to be laid to rest.

Ventilation:

Good ventilation will be your number two in terms of odor control. Part of keeping the bedding dry involves allowing air to flow and dry up the moisture. Fans are a poultry farmer's best friend; if set up right, they could be used to move the stale air and spread moisture by setting up misters to them.

Water:

Water is an essential element in the living period. Especially on hot days, your birds will need plenty of fresh water. Not only will water keep your birds alive and healthy, but without enough of it, later, you'll limit your egg production. A thirsty bird is an unnecessary headache.

Feed:

Feed and water are two peas in the same pod, the rule of thumb I employ for new chicks are to always keep food available during the day for egg layers and all night for meat birds. There is some math you can do to figure out the correct ratios, but to be honest, for what this is, it's better to learn what you need as you go.

Lighting:

Not to be confused with a heat source, but it can also be a heat source, which is super important. I said it all crazy like that because the truth is they can be two separate components. But if you were listening to me, I say to use white heat lamps for daytime and red/ blue heat lamps at night.

The reason you would need lights at night is to coerce the meat birds to continue to eat into the night and gain all that good fat. Birds naturally look to roost (sleep for the night) when the sun starts to go down. They know their limitations and don't pretend to be tough enough for that wild nightlife. So generally, when you have birds, you can let them loose during the day, and at night they will come back home to roost where you provide their safety.

.

4th Joint: 10|16 160 Self Sustainability: Grow your own feed for your food

This joint is dedicated to realism-ism. The entire point of this book is self-sustainability. To be self-sustainable, what you're doing has to be easily replicated, maintainable, cost-efficient, and most importantly, it must be self-serving. This means everything must be of value to the greater goal, and without one aspect, the whole thing is entirely different. When I think about this endeavor, I think about how much money it costs to feed a chicken. While a bag of feed on its own is not a ridiculous amount of money, the name of the game is to preserve as much of your resources as possible to see more tomorrows.

Chickens need to eat; Fish need to eat; plants need to eat; all these things need to eat before you can eat. Think about that for a second; Your food requires food. Not only does it need food, but it needs food consistently to be of value. The six months it takes for a chick to turn into a laying hen, that entire time, you will need to feed that bird. The 6 – 8 weeks of growing out your meat birds before you harvest them, during that time, your birds will need to eat a lot.

Not to be too dramatic, but these are the realities it costs to eat. Nothing in this world is free; we live in a society where unless you have a job, you cannot eat, and you're not even allowed to take food out of the trash bin of a restaurant. So yes, it costs to eat, but how you deal with that is the difference. There are a few things that I will lay out that have worked for me and from what I see from other farmers for them as well.

Fodder:

Grow your grass! I know it sounds a little silly, but when you delve into this world, you'll realize just how vital grass can be for livestock and humans. It takes about seven days for a seed to be fully sprouted and ready to be planted; at the same time, the sprout itself is at its' most beneficial for consumption; in the spirit of sustainability, you could grow a tray of seed out to be row planted to maturity for re-seeding.

10|16 160   82

So, we all "know" that grass-fed livestock is "healthier" than livestock that feeds on other animals. With that in mind, if you allowed your chickens to graze solely, they would be entirely free for you to sustain. The fauna and flora they'd dig up would suffice, but this is not the case for most people, so I'd suggest growing your own. Fodder, defined by a basic google search, is food, especially dried hay or feed, for cattle and other livestock. With this definition, you would think they are talking about different food from the food humans eat, but when you get down to it, they are talking about things like Wheat grain, Barley, Oats, etc. while I admit these foods are typical livestock feed, they are also good people feed as well.

You must be mindful of buzzwords when sourcing your seeds, but only in terms of what you'll pay for the product. I'm not here to try and scare you into doing something, but, for example, if you wanted to grow "organic" black beans, you could search online and go to an "organic" website claiming to be from some rainforest somewhere, or special hill and purchase your seeds; or you could go to your local grocery store and pick up a bag of any number dry bean soup and sprout those.

Micro Greens:

The business of micro-greens is another benefit to growing your grass. These greens are harvested very soon after they sprout from the seed, but I guess, being politically correct, these techniques wouldn't be called sprouts because when you harvest the greens, you cut at the soil line. Cutting at the soil line allows for regrowth making it a profitable endeavor.

They are nutritional supplements and can be used in a variety of ways. Because they are packed with so much goodness, some would consider greens at this stage to be "superfoods." If you were to do a deep dive into researching micro greens, you would find that with almost any green you choose, at this stage, most plants are rich in potassium, zinc, magnesium, copper, and iron, to name a few.

Fodder Rack System components/build

So how do you set up a fodder rack system? It's super simple. All you need is a set of trays and a rack to hold the trays on. You'll need a bucket/reservoir, a small pond pump, and tubing. You first determine your tray size; you need to know how much fodder your chickens will need to consume daily. On average, one of my birds will eat about 2% of its body weight on either commercial feed or organic grass before it doesn't want to eat anymore.

I did this food ratio test by trial and error; first, I isolated a bird and put it in a separate dog cage; I weighed it to see the differences from day to day based on its starting weight. I then took a bowl and filled it with chicken feed, weighed the bowl, and allowed the chicken to eat until it was satiated. After doing this for a week, I noticed that by reducing the amount of feed I put in the bowl, I figured out that a 50 lb bag of feed would feed 33 chickens well each day. In comparison, a 35lb bucket of wheat for sprouting will cost about $100, but don't lose yourself. If you were to purchase a bag of horse feed, you'd find that it's probably oats, wheat, and other grains; and will most likely sprout and it'll cost about $25 for a 50lb bag.

After you've determined how much grain you need, you'll then start the actual process. Soak your grains for at least 4 hours, but not longer than 24 hours for most grains. This process opens the grain up and lets it know it's time to be useful. Drain the grains and line them in your trays. Now your trays need to be stacked in a zigzag on the rack that allows gravity to force the water pumped from the reservoir to the highest tray in your system to flow downward into each tray below it. Holes are drilled in the front of each tray so the water can drip/drain from the top row down to each tray until it returns to the reservoir. You only run the pump a couple of times a day; if you set it correctly, you'll be putting out a new tray on the rack every day until you have enough for the week. The goal is that once you set the top tray, by the time the bottom tray is being placed, the top tray is ready to be fed to your birds, and you keep doing the process repeatedly.

Nothing about farming is easy and cheap. If it is easy, it is an expensive setup manufactured by engineers. It will require a lot more hands-on or sweat equity.

Another inexpensive and very sustainable way to feed the food on your farm to feed your family is to investigate grubs. Remember how I said if you allowed your birds to forage, they'd be able to fend for themselves? Keeping this in mind, one of their primary protein sources is the bugs they dig up. Don't forget about your fish; any fisherman will tell you fish love bugs.

So, this might be better or more useful information for people with a bit more space but consider this; you can get/build a composter that doubles down as a grub collector. Yes, "grub collector," I'm saying we grow our bugs too. This book is not endorsing a particular product, mainly because I'm not getting compensated for it. I would suggest a Protapod grub composter. I say this because of its design; it is configured in a way where your grubs, once they have eaten enough and moved away from the food, cannot go anywhere but into a hole where a bucket catches them with a tight lid on top. Your birds will love you for this. Your wallet will thank you for this. You fill it with food scraps; it doesn't matter how fresh, flies will do what flies do, and that's it.

The benefit of composting in this manner is two-fold; I hope you are getting a trend to how I think things need to be able to be hit from multiple angles or have various uses. Use # 1 is spelled out above but let's consider the other benefit of composting. The amount of natural organic fertilizer you'll get for your plants will be tremendous.

AQUATICS:

Minnows are an excellent addition to your setup; these little guys are tough and significant because they reproduce in a stock pond well enough to be used as a feed source for multiple animals. Minnows are also suitable for controlling mosquitos and insects in your pond or tank. My idea for these fish is to be harvested, grinded up, and turned into chicken feed, dog food, and other fish feed.

Tilapia is another staple I'm using in growing my feed for my feed. Like minnows, tilapia are tough fish, meaning you don't have to be that great in their husbandry to win still. My idea for these fish is very similar to the minnows; it takes between 9 -12 months for tilapia to go from fry (baby fish) to frying pan on average. When you breed these fish on average, you are looking at a few hundred fry each time one fish spawns; one fish will produce enough for one family to consume alone; but if done correctly, you could kill two birds with one stone and grind those adults up and turn them into chicken feed, dog food, and other fish feed.

Duck Weed is the grandest of all the grand ideas because this plant grows on the pond surface; it is a natural fish food for your minnows and tilapia to survive on alone if you desire. You may have to consider a protein source if you feed greens alone, but your fish will live and grow (all be it slowly) on this plant as its food source. Duckweed is so prolific that it sells by the teaspoon. You don't need a gallon of it; you only need to let it establish itself in your tank, and it will bloom.

You harvest the duck weed by scooping what you need from the surface of your tank/pond and **distributing it among your poultry.** Chickens fight something awful over the stuff. The only negative about duckweed is that when starting it, it requires a specific conditioning; the water hitting the tank/pond surface cannot be too turbulent; there is a happy medium about the flow rate of the falling water and what the plant likes. Typically, a slow to moderate flow will be fine, but if the flow is too much, the plant won't be able to situate itself, and it will stress out and die.

5th Joint: 10|16 160 Self Sustainability: The god Ali Final thoughts

It's easy to point fingers at someone other than yourself and assume that because of someone else's interference or sabotage, you are without a way to sustain yourself. In some instances, this may be very true; but in most situations in this world that we live in, not too many people truly have the resources to single you out specifically. In the minority of times, it's easy to see that you're probably witnessing bullying in action. If it's one thing I cannot tolerate too well, it's bullying, but the worse is the person allowing themselves to be bullied. Waiting for someone to pop up out of nowhere and care more about your life than you do yourself is not wise. Not to sound pious, but if you are not willing to do whatever you can to support yourself and your interests, you don't deserve happiness. It might be my schema, having survived the late 70s, 80's, and seizing the opportunity to allow myself to experience the possibility of becoming my best self, that I think the way I do. To me, "possession" being 9/10th of the law only means that if someone can relieve you of your happiness, it was not meant for you. Granted, you only have 100% control of yourself and your actions, and you really cannot stop someone from attempting to harm you, but it falls on no one else but you to put forth the most effort in your endeavor to live.

You cannot always punch your bully in the face and make them stop, but just because that option is off the table doesn't mean there aren't plenty of ways to protect yourself. You must be willing to win and keep winning or risk your livelihood. We must break the shackle of fear. As I stated at the beginning of this journey, fear is a hell of a drug; using popular culture, most people have been conditioned to support a widespread opinion that usually goes against their own natural discernment. While on my travels, I stumbled on the path of Prince Hall Masonry while having been initiated, passed, and raised on the level, and I learned many great lessons. None more impactful than that of the words found in Psalms 133; in it, we are taught to seek indeed to find happiness and righteous pleasure in your brother or sisters' prosperity. We are encouraged not to be jealous but rather hope for the success of your kin. In this, the hope is that we realize that the crabs in the barrel mentality are one brought upon a people whose minds have been bound and must be accessible. We are to be reminded that crabs naturally flourish on beaches around the world and have never been seen attempting to stagnate each other progress.

Poverty is the catalyst for most fears, halting a person's willingness to step outside of a constructed comfort zone and attempt to be the change they need to see in the world. This book is titled 10|16 160 Self Sustainability, and in no way am I intending to limit what it means to be self-sustainable. There are a million ways to provide for yourself and your family; no one way is the way; I hope to increase your options with this text. It is my desire through farming that I can show you an option that will assist you and your family in maximizing your resources and making a better, safer way in this crazy upside-down world we live in. Jay z said, "people love the announcement, but what about the execution of it" (Kevin Hart interview 35:25 https://youtu.be/DO1MO7we-RQ)

This is a powerful quotable by Hov; to me, it speaks to how excited we get when we see winners and how we live vicariously through them as if we are winners too. I know to some, this may seem in the same vein as Psalms 133, but to me, it's complete corruption. In no way does another sweat equity and sacrifice benefit you if you are out of a position to step up when it is your time. We are not to wish we had what our brother and sister have; we should instead be happy that they were able to find a piece of happiness. That in and of itself should be enough to inspire us to look within ourselves and find out what we need to be happy. Too often, our excitement makes us move too fast, we get so caught up in the possibility that we don't prepare for the things that may stand in our way, and we react instead of responding.

When you are sitting with your thoughts or sharing them with your family, think about what preparations and procedures you need to do to write a better future. I hope this text's words are plain and profound enough to follow and share with non-growers. I endeavor that as many people grow their food as possible. In doing so, we will be open sustainable to cooperation. This journey has taught me that teams win more than anything else. No one person is an island, and while individual efforts are paramount, we will be doomed as a society without collaboration. Thank you for your time and attention, and I hope for your prosperity and success.

10|16 160

www.ingramcontent.com/pod-product-compliance
Lightning Source LLC
Chambersburg PA
CBHW051209090426

42740CB00021B/3439

*9 7989 85 1 9 0 9 9 1 *